THE WHOLESOME DOG BISCUIT

THE WHOLESOME DOG BISCUIT

a barker's dozen

Patricia Leslie, M.S.

Regent Press
Oakland, CA

ISBN 1-58790-104-8

Library of Congress Control Number: 2004096081

Dog Portraits: Patricia Leslie

Manufactured in the U.S.A.

Regent Press
6020–A Adeline Street
Oakland, CA 94608
www.regentpress.com

www.wholesomeoven.net

This book is dedicated to Sharon Callahan, with gratitude
for her extraordinary wisdom, spiritual insight, and compassion,
for her profoundly healing and nurturing flower essences,
and for encouraging me to bring this book into being.

And this book is also dedicated
to each of the dear canine companions of my life —
those who are now in spirit,
those who currently share my home,
and those yet to come.

TABLE OF CONTENTS

INTRODUCTION

Let's face it – dogs are not much impressed by cute food. I have never yet had one of my dog-kids look at a commercial biscuit and say, "Oh, wowf, it's shaped like a postman, how cute! Now I really want to eat it!" or, "Hey, carob frosting, cool!" Also, since dogs are virtually color-blind, all they really get out of brightly-colored treats are a lot of unhealthy additive dyes (the same dyes that are implicated in learning disorders in human children).

The things that actually impress dogs are smell and taste and texture and quantity. They look at a biscuit and say, "Hey, food! Put it in my mouth, quick!" And if you're lucky, they mumble "Thank you" through the crunching, slurping sounds. So I have developed a few very useful biscuit styles, intended to produce the maximum number of biscuits with the least amount of fiddling around.

I first started baking biscuits when Maeve was one year old, and had just been diagnosed with multiple food allergies (rice, soy, corn, potatoes). To complicate matters, Toby was about 10, and had had a lifelong wheat allergy. This was 1998, and a desperate search of pet-food specialty stores did not turn up a single biscuit that was entirely free of allergens. I was already an experienced baker, so I thought it was time to start experimenting.

Well, my home-baked biscuits turned out to be a hit with all of the dogs. I spent the next couple of years baking all the dog biscuits that we needed. Then the world caught up with me, and a new generation of healthy – even vegan – dog biscuits began appearing on the shelves. I freely admit that having those commercial alternatives available is sometimes a lifesaver, when I am just too busy or exhausted to enjoy baking up a batch. I am not of the school that says we must do everything the hard way all the time in order to appear worthy to our loved ones. Biscuit-making for your doggy should never become drudgery. (For a list of ready-bought natural treats, see page 52.)

These days I still try to home-bake most of the dogs' treats. For one thing, those "alternative" biscuits are in general ridiculously expensive. In the second place, all of them are really too big and bulky to make practical training treats. Some just don't break into small pieces easily. Others crumble. Also, even though they have healthy ingredients, I can't escape the suspicion that they are all a bit stale by the time I buy them. Finally, I have noticed that some theoretically healthy biscuits have ingredients which I consider questionable, such as raisins.

When I bake for the dogs, I know that they are getting the freshest human-grade ingredients, and no preservatives at all, even "benign" ones. I can insure that they are avoiding allergens. And the best reason of all is that the dogs love fresh biscuits. They always know when I am baking for them, and as the biscuits slowly harden in the oven, they make frequent trips to the kitchen, to check on the progress. By the time I am removing the fully-baked outer rows from the baking sheet, they have gathered with expressions of delighted expectancy. When I say, "Okay, time to test the new batch," their joy is complete.

So, my simplest and best reason for baking biscuits is that making dogs happy makes me happy. And this is only right, because I have no doubt that making their people happy is the greatest joy in the hearts of dogs. In fact, I believe that this reciprocity of happiness is the essence of the dog-human symbiosis. Surely, this is the real reason they threw in their lot with us all those millennia ago and chose domestication.

I've put this book together for all humans of like mind – those who love their dogs so much, that a little extra work on their behalf is a pleasure, rather than a chore. This book comes to you, dear reader, with my sincere wishes that you and your furry loved ones will enjoy many joyous and healthy years together. I trust these recipes will contribute to both the joy and the health.

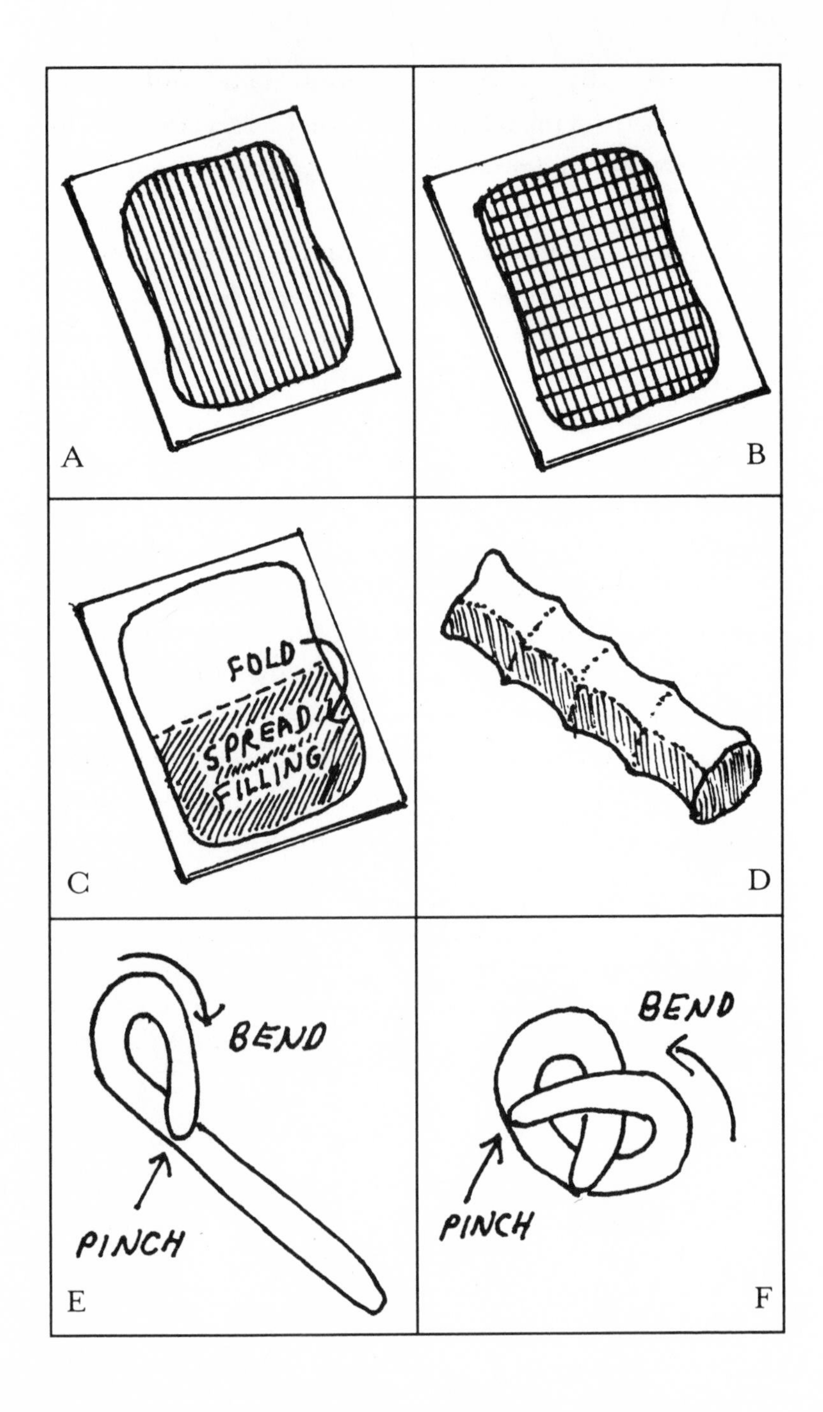
A
B
FOLD
SPREAD FILLING
C
D
BEND
PINCH
E
BEND
PINCH
F

PREPARATION TECHNIQUES

My basic biscuit is the thin, flat rectangle. I like to make them about 3/4 x 1 1/4 inch, and 1/4 inch thick at most. I always need lots of training treats, and these break easily into smaller fragments which go down fast during training. For simply expressing love and approval, a whole one is a satisfying size. Also, if your dog owns any of those sturdy rubber balls with slots bored through them, a biscuit this size is easy to press into the slots, and provides several minutes of chewing, chomping, crunching fun. For a bigger dog and more crunch, you can make Pretzel Cookies or "Bone-Sticks." "Dog Newtons" have extra sweet-tooth appeal thanks to a very thin layer of fruit preserves. You don't really need any more biscuit styles than these. However, if you happen to have acquired a bone-shaped biscuit cutter (or cat-shaped, or dogcatcher-shaped), you can try it out on rolled out dough.

The following instructions are general. Some recipes will give a variation in baking times.

Measuring

All flour measurements are for unsifted flour. Fruit and vegetable measurements are approximate. All these recipes are flexible, since the moisture content varies in fruits and vegetables. You can always add a bit of extra flour if the dough seems too wet to roll out.

Mixing

A food processor is essential to mixing up all the doughs except for Winnie's Quickie Biccies.

Baking sheets

The baking sheets I used for all these recipes are rectangular, 14 inches by 16 inches, and flat, with no raised rim. If your baking sheets are much smaller, either use more of them, or roll the dough much thicker.

Even if your baking sheets are nonstick, oil them generously, using a spray-on blend of canola oil and lecithin.

Flat biscuits: rolling

Scrape the dough into the middle of the oiled baking sheet (divide it equally if the recipe recommends two sheets). Have a saucer handy, with 2-3 tablespoons of canola oil poured onto it. Oil your palms well, and begin pressing and scrunching the dough down onto the sheet. Alternate between pressing down and out from the center in all directions, and smoothing, tucking and evening out the edges. Re-oil your palms as necessary.

When the dough is about 1/2 inch thick all over, switch to your rolling pin to get it thinner. Aim for a uniform

thickness. Roll from the center to the edges, changing directions as needed. If you don't have a rolling pin, just keep pressing the dough out by hand.

All these flat biscuits work best when they are between 1/8 and 1/4 inch thick. Any thinner than 1/8 inch and they crumble too easily. The ones that have larger particles (of oats or vegetables for example) will hold together best if they stay about 1/4 inch thick. (If you choose to roll it thicker, you should increase the baking time proportionately; but don't increase the temperature.)

Flat biscuits: cutting

After rolling out the dough, bake it for about 10-15 minutes. After this time, the surface of the dough will have a "skin" which is smooth and dry to touch – but just underneath the dough will still be very pliable. Remove the sheets one at a time to work on them. After cutting, return them to the oven.

You'll need something to cut the dough. I don't use anything with a very sharp or serrated edge on non-stick baking sheets. If you have a rolling pastry cutter, you can use that with or without a straight-edge guide. You might use a bread-dough chopper if you have one. What I use is a 5 inch long, rectangular metal pancake turner.

Rectangles: Start at one of the longer edges. Cut a strip about 3/4 inch in from the edge of the dough. Press your cutter straight down into the dough. Continue cutting the whole sheet of dough into strips about 3/4 inch wide (Diagram A). Then cut the strips crosswise, into sections about 1 1/4 inches long (Diagram B).

Cookie cutter option: Starting from one edge, press the cookie cutter straight down into the dough. Keep the cookie shapes as close together as possible. Leave them in place. There will be "wasted" dough in between the cut-out cookies. Use a small, blunt knife to cut these in-between pieces into sections about an inch long. You will end up with a number of cookie-cutter biscuits, and quite a few smaller, abstract-shaped treats. Tell your dogs the abstract ones are sea monsters or bugbears or Martians. They won't know the difference!

Bone-Sticks

These are easy and fun. Children can make these. In fact, anyone who has ever played with any kind of clay will know how to make these.

Start with a lump of dough about the size of a standard walnut. Scrunch it in your fist a few times, until it is free of air pockets. Then wrap all your fingers around it, and form a gently-closed fist. You will thus squeeze the dough into a horizontal, generally cylindrical shape. The impression from your fingers will create a slightly scalloped surface (Diagram D). Use the thumb and index fingers of your other hand to tuck and press the dough in a bit at each end.

Lay these in rows on the oiled baking sheet, about 1/2 inch apart. Press them down a little bit to flatten the underside, so that they don't roll off easily.

Dog Newtons

Work with one-third of the recipe's dough. (You can use the rest of the dough to make Bone-Sticks and/or Pretzel Cookies. Or, triple the amount of filling and repeat this process twice with the other 2/3 of the dough.)

Roll the portion of dough out to form a long rectangle about 12x15 inches, and about 1/8 inch thick. Place the dough rectangle on the baking sheet. Spread the oat filling evenly over one-half of the dough (Diagram C). Fold the uncovered half of the dough over the filled half. Seal the three open edges with a little water. Do not pre-bake, but otherwise follow the above directions for cutting rectangualr biscuits. Cut straight down through all the layers; do not separate the pieces.

Dog Newtons filling

(this is enough for 1/3 of the dough):

3 tablespoons quick oats

3 tablespoons fruit preserves* (berry, apricot, peach, rosehips, or applesauce)

Stir together the oats and fruit preserves to form a paste.

> * WARNING! PLEASE NOTE!
> NOT grape, grape-sweetened, or citrus preserves

Pretzel Cookies

These are not hard to make; however, they won't work well with the doughs that have bigger particles in them (like ground-up oats or vegetables).

Start with a lump of dough about the size of a standard walnut. Scrunch it in your fist a few times, until it is free of air pockets. Then roll it between your palms (like making a clay snake). Make a rope about 1/4 inch thick and between 2 1/2 and 4 inches long. Bend one end of the rope around to meet itself a little more than halfway to the other end. You will have a loop sort of like the letter "P" (Diagram E). Pinch the dough together where the rope meets itself. Then bend the other end around to overlap the first loop in an "x" (Diagram F). Pinch the dough to itself where this second end meets the rope.

Lay these in rows on the oiled baking sheet.

Oven Rack Placement

For one baking sheet, keep the oven rack halfway between the oven floor and ceiling. For two baking sheets, keep both near the center of the oven, about six inches apart.

Baking: Flat Biscuits

Always preheat the oven. Be sure to use the temperature recommended with the recipe. Fairly long baking times at a low temperature is the key to a hard, crunchy biscuit. It insures that the biscuits will completely dry out inside, without scorching or burning on the outside. The biscuits will keep for many weeks if you get them completely dry and hardened.

Place one baking sheet on each rack. Set the timer for 10 to 15 minutes. Then remove the sheet(s) from the oven one at a time to cut the biscuits (see above). Put the first sheet back before removing the second one.

After returning the second sheet to the oven, set the timer again. Bake the biscuits for about another hour (check the recipe for any baking time variations). When the timer goes off, remove the baking sheet(s) from the oven one at a time. Turn all the biscuits over. Remove

all the done, fully-hardened ones from the sheet(s). The biscuits along the edges will be the hardest. Test them by trying to dent the undersides with your thumbnail. If your nail goes in at all, they need more baking. If you can't make a dent, they are done. If you're not quite certain, break one. If it has a crispy "snap," it's done.

Return the sheet(s) with the turned-over biscuits to the oven for another 30 minutes (or whatever the recipe suggests). Check them, again removing all additional biscuits which have hardened up. Do not turn them back over. Return the sheet(s) to the oven for 15 minutes at a time, checking for doneness each time. When only a few dozen biscuits show any "give" when pressed with a fingernail, turn the oven off, leave it shut, and let those last biscuits "coast" until the oven is cool.

Please note: all the biscuits generally feel a bit soft as long as they are warm, and harden as they cool, especially the ones made with molasses. So give them a minute or two out of the oven when checking for doneness.

Baking: Dog Newtons

Do not pre-bake the dough before cutting it into rectangles. Set the timer for the recommended baking time according to the recipe, *plus* 10 minutes. When the timer goes off, turn all the biscuits over.

Check them for doneness as above, and finish baking as above.

Baking: Bone-Sticks and Pretzel Cookies

Set the timer for the recommended baking time according to the recipe, *plus* 10 minutes. When the timer goes off, remove the baking sheets from the oven one at a time. Turn all the biscuits over. Check them for doneness by trying to dent the undersides with your thumbnail. Remove any that you find are fully-hardened.

Return the others to the oven. Check them for doneness after another 30 minutes, and every 15 minutes after that. Each time you check on them, remove the ones that have become fully-hardened. Test the doneness by breaking one open. It should snap crisply.

With the bone sticks, turn them over as best you can. Since they are somewhat cylindrical, they won't necessarily turn 180°. The important thing is to get the undersides up off the baking sheet so that they will harden faster.

RECIPES

ANDREW'S GREAT DANISH COOKIES

Makes about 1 1/4 pounds of biscuits

These bake very hard and crunchable, which makes them a good choice for larger dogs, and it is easy to make them as big as you need. All these features make this a good choice for a very large dog who needs a treat that won't get lost in his mouth. This very workable dough is perfect for kids to use. They can create "Dog Newtons," "Bone-Sticks," and Pretzel Cookies.

Preheat oven to 275°.
Oil one 14x16 inch baking sheet.
For "Dog Newtons," have your rolling board floured with extra spelt flour.

2 cups barley flour
1 1/4 cups spelt flour

1/3 cup almond butter
3/4 cup oat milk
1 tablespoon maple syrup
1 tablespoon canola oil

1/4 cup cold water, as necessary

In a food processor, process together the almond butter, oat milk, maple syrup and oil.
Add the barley flour to the mixture, processing everything together well.
Add the spelt flour and process everything again.
If the nut butter was very oily and runny, you may not need much water, if any. However, if the mixture turns out crumbly, then add the cold water through the feeder tube while running the processor, until a cohesive ball of dough forms. The dough should be very stiff and workable. If it is too sticky to work with, chill it for 20 minutes.

Prepare Dog Newtons, Pretzel Cookies, or Bone-Sticks, according to the instructions on pages 16-18.
Bake according to the instructions on pages 18-20.

BRUCE'S SCOTTISH OATCAKES

Makes about 1 1/4 pounds of biscuits

This is the simplest recipe of all — almost 100% oats. For dogs with many dietary issues, these might be the best choice. Oats are highly digestible for dogs, and they all seem to love the flavor. These crumble easily.

Preheat oven to 325°.
Oil two 14x16 inch baking sheets.
Have your rolling board ready.

1/3 cup oat bran or rolled oats*

4 1/2 cups rolled oats
1/2 teaspoon sea salt

3 tablespoons canola oil

1 1/8 cups hot water

Run the 4 1/2 cups of oats with the salt in the food processor until they are mostly coarse-ground, with 10 to 20 percent of the oats still in whole or partial form. Pour the oats into a large mixing bowl.
Drizzle the oil over the oats. Stir everything by hand until all the oats are slightly moistened and clumping a little. Pour in the hot water. Stir it in thoroughly. A stiff dough will form.

Divide the dough into two equal portions; work with one at a time. Knead and press each one into a cohesive sphere, then into a flattened disk.

Sprinkle half of the bran or flour onto your rolling board. Press all sides of the dough disk into the bran or flour, coating it. Flatten it more with your hands. Roll it out to 1/4 inch thickness. Roll mostly in one direction, to get an oval or rectangular piece of dough approximately 16 inches by 10 inches.

Cut the dough into inch-wide strips. Cut the strips into sections 2 or 3 inches long. Use a spatula to transfer the strips to a baking sheet. Leave very narrow spaces around them.

Bake for 40-45 minutes. They will be quite firm and dry. Some may arch upwards. They do not need to be turned over. Check doneness by breaking one. It should snap crisply, and feel dry inside.

**If you don't have oat bran on hand, grind an extra 1/3 cup of oats into flour in the food processor. Reserve this oat flour.*

COUSIN BEATRICE'S LITTLE SWEETIES

Makes about 1 1/2 pounds

Carob is the safe way to satisfy a dog's interest in things chocolate.

Oil two 14x16 inch baking sheets.

5 cups rolled oats, ground to flour (about 4 1/2 cups of flour)
1/4 cup carob powder*

1 1/2 cups (about 3) ripe bananas
1/2 cup applesauce
OR 1 more banana
2 tablespoons canola oil

2 cups cooked barley

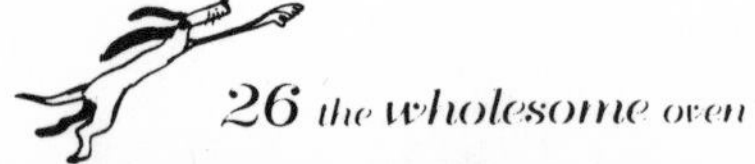

If you don't have leftover cooked barley, cook it first. One cup of dry barley flakes makes about 2 cups cooked.

Grind the oats to flour in a food processor. Pour the oat flour into a bowl.
Stir the carob powder into the oat flour.

In the food processor, process the bananas, applesauce and oil together until well pureed.
Add the cooked barley and process everything again until well combined.

Gradually add the oat flour until a stiff dough forms. This dough will be quite sticky. Scrape the dough into the bowl that you used for the oat flour, and chill it for about 30 minutes.

MEANWHILE, preheat oven to 250°.

Prepare the biscuits according to the instructions on pages 14-15. (Roll these out about 1/4 inch thick.)
Bake according to the instructions on pages 18-19.

***WARNING! PLEASE NOTE!**

DO NOT SUBSTITUTE COCOA for the carob. See page 56.
Use only carob, or simply add 1/4 cup of barley flour.

JEFF'S HIGH ENERGY SNAPS

Makes about 1 1/4 pounds of biscuits

Jeff is prone to zooming around the yard at high speed for the sheer joy of it. These work best as flat biscuits.

Preheat oven to 250°.
Oil two 14x16 inch large baking sheets.

1/2 cup dates*, pitted

1 1/2 cups cooked grain
(barley, millet, oats, rice, quinoa, amaranth, triticale, or a combination)
1/2 cup raw almond butter
1/2 cup oat or nut milk
2 tablespoons honey or maple syrup
2 tablespoons canola oil
1/2 teaspoon vanilla (optional)

1 1/2 cups rolled oats
1/4 cup carob powder*

2 – 2 1/4 cups barley flour, unsifted

***WARNING! PLEASE NOTE!**

DO NOT SUBSTITUTE RAISINS for the dates. **See page 54.**

Dried blueberries, pre-soaked, will work.

DO NOT SUBSTITUTE COCOA for the carob. **See page 56.**

Use only carob, or simply add 1/4 cup of barley flour.

Use a food processor to grind the dates.
Add the cooked grain, almond butter, oat milk, sweetener, oil (and vanilla).
Process all these ingredients until they are fully blended.

Stir the carob powder into the oats and add them to the mixture in the processor.
Process the oats into the date mixture just until everything is well-mixed, but not pureed. There should still be many small bits of oat visible.
Scrape the dough into a large bowl. Use a wooden spoon to fold in the barley flour by hand, until the ingredients form a fairly stiff dough.

Prepare the biscuits according to the instructions on pages 14-15.
Bake according to the instructions on pages 18-19.

MAEVE'S FAVES

Makes about 1 1/8 pound of biscuits

Maeve adores pears. She not only gleans the overripe ones from the ground under the tree, she taught herself to "harvest" them by jumping up and knocking them down with her nose! This dough will make Bone-Sticks, but is too soft and sticky for Pretzels.

Preheat oven to 250°.
Oil one or two 14x16 inch baking sheets.

2 cups (2 or 3) pears, peeled, cored and chopped (raw or canned)
OR
2 cups (2 or 3) raw apples, peeled, cored and chopped
OR
1 cup applesauce
1/4 cup pear nectar or apple juice
2 tablespoons canola oil
1 teaspoon vanilla extract (optional)

3/4 cup oat bran, quinoa flakes or quinoa flour

1 3/4 cups millet flour, unsifted
1 to 1 3/4 cups buckwheat flour, unsifted

Use a food processor to process the fruit, juice, oil (and vanilla) together until everything is well-blended. The fruit should be reduced to mush.

Add the oat bran or quinoa and process until everything is combined.
Add the millet flour and process until everything is fully combined.
Add the buckwheat flour 1/2 cup at a time, processing each time until it is fully blended in. The dough should become workable with the addition of the buckwheat.

Prepare the biscuits according to the instructions on pages 14-15. (Roll these out about 1/8 inch thick.)
Bake according to the instructions on pages 18-19. (Bake for about 40 minutes before turning them over.)

NEIGHBORLY NIBBLES

Makes about 1 1/3 pounds of biscuits

This recipe is dedicated to all the beautiful and delightful dogs whom I have come to know as friends over the years — through my family, my neighbors, training classes, the dog park, and fostering: King, Blackie, Rufus, Darcy, Gizmo, Sutter, Abby (both of them), Sam, Misha, Iago, Buster, Winston, Bonnie, Lady, Rudder, Tasha, Maddy, Bella, Yulee, Taz, Oso, Princess, Tucker, Chuy, Molly (all 3 of them), Rosie, True, Slick, Mick, Blue, Farley, Sally, Sydney, Markie, Lady Guinevere, Pepsi, Brutus, Ellie, Dusty, Boomer, Bandit, Harriet, Girldog, Honey, Daisy, Sheela, Penny, Bryce, Tilly, Murphy, Kokoro, and last but certainly not least, Jack the Dog, who left Animal Control to have a happy life with my Aunt Ann, Uncle Vic, and Cousin Ken.

Preheat oven to 250°.
Oil two 14x16 inch baking sheets.

1 large (about 1 1/2 cups) sweet potato, cooked*
3/8 cup applesauce
1/4 cup oat milk
2 tablespoons canola oil

1 1/2 cups oat bran
1 1/2 to 2 cups barley flour

Combine the cubed, cooked sweet potato, applesauce, oat milk and oil in a food processor. Process these ingredients together until well-blended.

Add the oat bran and process the mixture until it is well-combined.
Add a cup of the barley flour and process the mixture again until it is well-combined.
Add another 1/2 cup of barley flour, and process the mixture until it forms a stiff dough. If it is still too soft and sticky to work with, add up to another half cup of barley flour and process again.

Prepare the biscuits according to the instructions on pages 14-15. (Roll these out about 1/8 inch thick.)
Bake according to the instructions on pages 18-19.

*If you happen to have a leftover baked sweet potato, peel it and cube it. Otherwise, cook it as follows. Peel it, cut it into 1 inch cubes, and place it in a saucepan with enough water to cover it completely. Bring the sweet potato to a boil, reduce the heat to medium, and continue boiling it until it the cubes are cooked through (about 15 minutes). Check by poking a few pieces with a knife. Remove the pan from the heat and drain off the water.

PAUL'S TRICKY TREATS

Makes about 1 1/4 pounds of biscuits

Preheat oven to 250°.
Oil two 14x16 inch baking sheets.

5 cups rolled oats, ground to flour

2 cups barley cereal or barley flakes, cooked
1 1/2 cups pumpkin puree
2 tablespoons canola oil
2 tablespoons molasses
1-2 teaspoons powdered ginger

up to 1 cup barley flour, as needed

Grind the oats to flour in the food processor. Transfer the oat flour to a bowl.

Combine the cooked barley, pumpkin, oil, molasses and ginger in the food processor. Process these ingredients together until well-blended.

Gradually add the oat flour to the mixture in the food processor, a cup at a time, until a stiff dough forms. If you use it all up and the dough is still quite wet and sticky, add up to another whole cup of barley flour, as needed. (If the food processor gets too full, transfer all the ingredients to a large bowl and finish working in the flour by wooden spoon and hand.) The dough should reach a point of feeling quite stiff and workable.

Prepare the biscuits according to the instructions on pages 14-15. (Roll these out about 1/4 inch thick.)
Bake according to the instructions on pages 18-19.

PERDIE'S TAIL WAGGERS

Makes about 1 pound of biscuits

Preheat the oven to 250°.
Oil two 14x16 inch baking sheets.

1 ripe avocado, pitted and peeled
1 cup peas, canned or frozen & thawed
1/4 cup water
2 tablespoons brewers yeast
1 1/2 tablespoons canola oil
1-2 teaspoons kelp granules

2 cups oat flour
1 cup millet flour

Put the avocado, peas, water, yeast, oil and kelp in a food processor. Process everything together until fully combined.

Add the millet flour and process it until well-combined. Add one cup of the oat flour. Process everything again. Add the second cup of oat flour. Process everything again. The dough should be quite stiff and workable.

Prepare the biscuits according to the instructions on pages 14-15. (Roll these out about 1/4 inch thick.)
Bake them according to the instructions on pages 18-19.

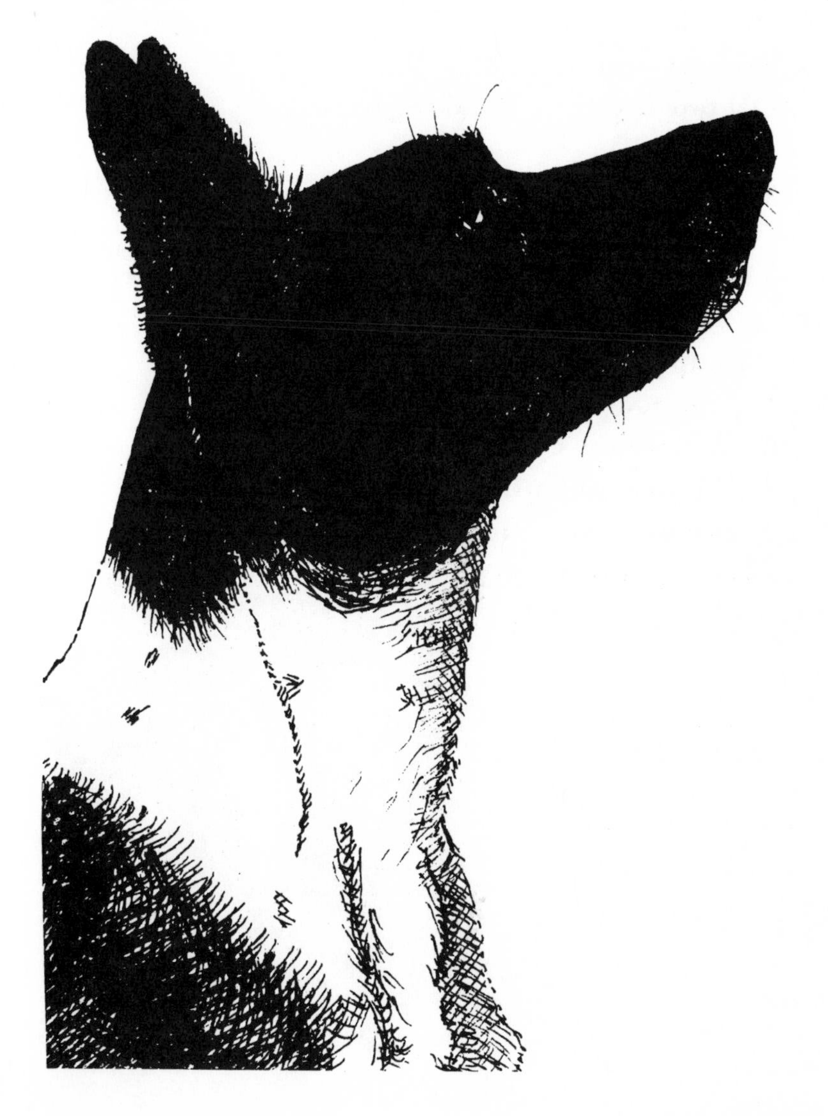

SENECA'S VEGGIE CRUNCHERS

Makes about 1 1/2 pounds of biscuits

Dogs are not really carnivores, but omnivores. Most dogs learn to love vegetables if they get the chance to try them.

Preheat oven to 275°.
Oil two 14x16 inch baking sheets.

3 3/4 cups rolled oats, ground to coarse flour

1 1/2 cups vegetables,* cut into 1/2 inch pieces

1/2 cup apple juice
1/3 cup water from the vegetables
3 tablespoons canola oil
3 tablespoons nutritional yeast
2 teaspoons kelp flakes
1/4 teaspoon turmeric**

1/2 cup oat bran

Put the vegetables in a saucepan with just enough water to cover them. Boil them until they are completely soft (usually about 20-25 minutes). Add more water as necessary to keep them barely covered.

Grind the oats to coarse flour in a food processor. Pour the oat flour out into a bowl.

In the food processor, combine the cooked vegetables, apple juice, water, oil, yeast, kelp and turmeric. Process these together until they are well combined.

Add the oat flour about a cup at a time, processing the mixture well each time. Stop as necessary to scrape the dough down off the sides of the processor.

Add the oat bran. If the processor is already too full, scrape the dough out into the bowl with the rest of the oat flour. Then use a wooden spoon to mix the flour and bran into the dough. The dough should be stiff and fairly dry-feeling.

Prepare the biscuits according to the instructions on pages 14-15. (Roll these out about 1/4 inch thick.)
Bake them according to the instructions on pages 18-19.

***For a complete list of vegetables, see page 58.**
****Turmeric is not harmful to dogs. See page 58.**

TAMA'S ISLAND TARO SNACKS

Makes about 3/4 pound of biscuits

Taro, generally in the form of poi, was a staple food for dogs as well as people throughout traditional Polynesia. It is a very nutritious form of starch, and when you work with it, you the cohesiveness of the dough will show you just how starchy.

Preheat the oven to 275°.
Oil one 14x16 inch baking sheet.

1 1/2 cups (about 3/4 lb.) taro root
1 small banana, sliced
1/4 cup water
2 tablespoons canola oil
1 teaspoon powdered ginger
1/2 teaspoon kelp flakes
1/2 teaspoon sea salt

1 cup brown rice flour
1 to 1 1/3 cups millet flour

Put the taro roots, unpeeled, in a pot with 6 cups of water. Cover the pot and bring the water to a vigorous boil (about 8 minutes). Keep the pot covered and continue boiling vigorously for another 25-30 minutes, or until a knife goes all the way through the roots easily, like a boiled potato. Peel off the skins. Cube the roots.

Combine the taro, banana, water, oil, ginger, kelp and salt and in your food processor. Process all these ingredients together until they are well-blended.

Add the brown rice flour. Process the mixture until all the ingredients are pretty well combined.
Add 1 cup of millet flour. Process everything until it is fully combined and forms a very cohesive dough ball. If the dough seems too sticky, add several more tablespoons of millet flour. Add more flour as necessary. Process after each addition. (The dough will always feel a little soft and sticky.)

This dough is too sticky for a rolling pin. Turn the dough out onto the middle of your baking sheet.
Have a saucer of canola oil handy, or your can of spray-on canola and lecithin. Oil your palms (or spray the surface of the dough). Press and pat the dough out as evenly as possible, until it is about 1/4 inch thick everywhere, and covers most of the baking sheet. (Leaving about a half-inch margin uncovered.)

Bake according to the instructions on pages 18-19. (Let them bake for 15 minutes before cutting the dough into biscuits.)

TOBY'S DELIGHT CARROTY BITES

Makes about 3/4 pound of biscuits

Once we caught Toby in the vegetable garden, digging up baby carrots and sharing them with Andrew.

Preheat oven to 300°.
Oil one 14x16 inch baking sheet.

1 cup carrots, cut into 1/2 inch pieces
2 1/2 cups oat flour (approximately)
OR
3 cups rolled oats

1/2 cup carrot juice
3 tablespoons canola oil
2 tablespoons molasses

1/2 to 1 cup barley flour

Put the carrots in a saucepan with just enough water to cover them. Boil them until they are completely soft. Add more water as necessary to keep the carrots barely covered.

If using rolled oats, grind the oats to flour in a food processor. Pour the flour out into a bowl.

Process the cooked carrots, carrot juice, oil and molasses together in the food processor until they are fully combined.

Add 1/2 cup of barley flour and process it with the carrot mixture until the ingredients are fully combined.
Add the oat flour one cup at a time, processing well each time. You will probably have to stop it and scrape the dough down several times. If the dough is a little too moist, add another 1/2 cup of barley flour. This dough will probably not form a ball, but it will become a stiff, fairly dry-feeling dough.

Prepare the biscuits according to the instructions on pages 14-15. (Roll these out to about 1/4 inch thick.)
Bake according to the instructions on pages 18-19. (Shorten the time to 40 minutes after cutting them, and 30 minutes after turning them over.)

URSULA'S LANGUES DE CHOW

Makes about 1 1/2 pounds of biscuits

Ursula is not pure chow-chow; she's mixed with golden retriever. The result is a beautiful lavender-blue tongue. The blueberries make these biscuits come out very nearly the same color. (Langue is French for tongue.)

Preheat the oven to 250°.
Oil two 14x16 inch large baking sheets.

3/4 cup boiling water
1/2 cup dried blueberries*

2 cups cooked oatmeal
1 – 2 tablespoons kelp granules

1 cup millet flour
1 cup oat bran
OR
quinoa flakes

about 2 cups
barley flour

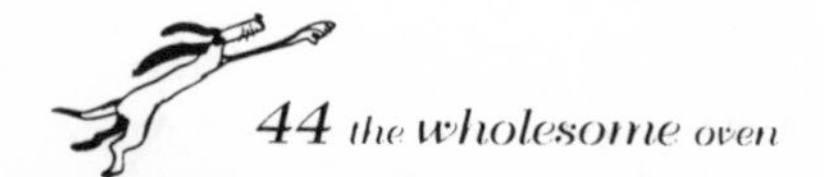

Pour the boiling water over the blueberries and let them sit for at least 15 minutes to soften.

Put the cooked oatmeal and kelp granules in a food processor. Add the soaked blueberries and any unabsorbed water. Process these ingredients together until well-combined.

Add the millet flour and process the mixture again.
Add the oat bran or quinoa flakes and process the mixture again.
Add about 1/2 cup of the barley flour and process the mixture again, until all the ingredients are well-combined.

Turn the mixture out into a large bowl. Fold or mash in the rest of the barley flour. Start with a wooden spoon. You may want to switch to using your hands.

Prepare the biscuits according to the instructions on pages 14-15. (Roll these out about 1/8 inch thick.)
Bake according to the instructions on pages 18-19.

*** WARNING! PLEASE NOTE!**

DO NOT SUBSTITUTE RAISINS for the blueberries.

See page 54.

WINNIE'S QUICKIE BIGGIES

Makes about 1 1/4 pounds of biscuits

These bake much faster than most of the other recipes, so they're handy when you can't spend 2–3 hours hanging around the kitchen.

Preheat oven to 325°.
Oil one 14x16 inch baking sheet.

2 1/2 cups barley flour
3/8 cup soymilk powder
1 teaspoon powdered ginger

3/4 cup molasses
1/4 cup canola oil

Sift together the barley flour, soy powder and ginger. Make a "well" in the middle of the mixture.

Warm the molasses and oil together. Stir them together occasionally. When the mixture is quite warm, pour it into the barley flour mixture. Use a wooden spoon to stir all the ingredients together until they are well-combined. The dough will be quite stiff and cohesive.

Prepare the biscuits according to the directions on pages 14-15. (Roll these out about 1/8 inch thick; the dough will nearly cover the whole baking sheet.)

Bake these biscuits for only 5 minutes before cutting the dough into biscuits.
Bake for another 15 minutes.
Turn all the biscuits over, removing any edge pieces that are completely hard.
Bake another 5 minutes. Check and remove additional hardened biscuits.
Bake another 5 minutes. Remove the remaining biscuits from the oven. They will harden as they cool.

Please note: These biscuits, because of the molasses, will feel very soft while warm, and get much harder as they cool, so be sure to remove some pieces from the hot baking sheet and give them a minute before testing them for hardness. A clue to doneness is that they turn a rich, dark brown.

FOOD, HEALTH AND DOGS

WHEN YOU DON'T HAVE TIME TO BAKE

My husband and I have been committed to training our dogs ever since Toby came into our lives in 1988. We had already been vegetarian for many years. So right from the beginning we have had the problem of not wanting to even touch, let alone purchase, most of what is sold as dog biscuits and training treats. You, too, may wish your dog to be a vegetarian or vegan for reasons of ethics, health, or both. But remember, dogs don't get the point. Genetically, canids are opportunistic scavengers, and omnivores. You will never convince them that decaying bits of dead animal are not desirable.

Modern, progressive dog training is reward-based — as opposed to the old-fashioned, harsh, correction-based method, which is not very effective. If you are in a dog-training class with a progressive-minded trainer (and why would you want to take your dog to the other kind?), you will need to employ generous amounts of tasty and high-quality treats. Undoubtedly, you will be encouraged to use treats that are based on processed meat by-products. Or your trainer may work with your dog, and reward with whatever is in her treat bag. For that matter, most training facilities' floors — and dog-park grounds — are covered with dropped bits of treat anyway. And some dogs are clearly part vacuum cleaner.

At some point you must make a choice: either allow

your dog to take a few calculated dietary risks so that she can fully enjoy her doggy life, or install her in a huge environment-controlled version of a plastic "hamster ball." My personal approach is to make every reasonable effort to control my dogs' diets, but not sweat it if something un-vegetarian gets in my dogs' mouths by either opportunism or well-meaning human donation.

So I recommend that if the trainer wants to give your dog something like freeze-dried liver, don't get into a debate about vegan philosophy. Dog-training classes are short on time, and everybody is there to get the maximum activity for their buck. Holding forth on your personal philosophy while the dogs get restless – and the people get frustrated – is not the way to win future converts to veganism. The tactful approach is to simply mention to the trainer each week that your dog is allergic to a wide range of foods, and that you will provide hypo-allergenic treats whenever necessary. (But make sure you do indeed come well-supplied.)

You can always use your dog's usual brand of kibble, and the usual mealtime amount, as your training treats. For a class, be sure to not feed the previous meal. For home training, make meal time your training time. This really gets dogs to "sit up and take notice" – when the only way to get dinner is to work with you!

These days, nobody has time to always bake all the biscuits they need. So here is my personal list of "convenience" snacks and treats, for those times when the biscuit cupboard is bare, and you had no intention of fetching your poor dog a disgusting old bone anyway.

For training treats and rewards in general:

Apples and pears, fresh, cut in 1/2 inch cubes
Banana chips, broken (go easy if they contain sugar)
Bananas, soft-dried, cut in 1/2 inch pieces
Bell peppers, raw, seeded and cut in 1/2 inch cubes
Blueberries, fresh or dried
Blackberries, fresh or dried
Cabbage core, lightly steamed, cut in 1/2 inch cubes
Carrots, lightly steamed, cut in 1/2 inch cubes
Cereal, dry (free of wheat, corn, sugar or food coloring)
Dates, cut into 1/2 inch pieces
Lettuce, torn in 1 inch squares
Rye crackers, broken (plain, not spicy)
Strawberries, dried
Taro chips, broken (plain, not spicy)
Vegan cheese, cut into 1/2 inch cubes
Veggie commercial kibble (a different flavor or brand than for meals)
Veggie commercial dog biscuits, broken into bits

If your dog tolerates corn and rice, also try:

Cereals, dry (any grain, free of sugar or food coloring)
Corn chips (plain, not spicy)
"Puffed" snacks (plain, not spicy)
Rice cakes (plain, not spicy)

And for basic chewing pleasure:

Edible:

babies' teething biscuits, breadsticks, rusks,
cabbage cores (whole, raw), carrots (whole, raw)

Non-edible:

Nylon bones (these come in different hardnesses, so you should keep an eye on your dog at first to see how fast the "bone" gets chewed into bits; if it gets whittled down a lot within 30 minutes, take it away and get a harder type)

Kongs® (by themselves, or stuffed with snacks)

Rubber "biscuit" balls with slots or holes to push biscuits inside (these should also be made of hard, resistant, non-crumbling rubber)

Knotted rope chew-toys (these need supervision)

> Be aware that even nylon bones and hard rubber toys can cause tooth breakage, if the dog is an obsessive chewer.

THINGS NOT TO FEED DOGS EVER

These substances:
alcoholic beverages
chocolate
BHA, BHT, ethoxyquin, MSG, potassium sorbate,
propyl gallate, propylene glycol, sodium nitrite
refined sugar (sucrose)
artificial flavorings
food colorings
"Meat"*

These plants and plant parts:
Apple seeds, leaves, stems
Avocado seeds
Citrus fruits
Eggplant stems
Fava beans
Grapes
Onions
Pineapple
Raisins
Raw potatoes & stems
Spices which are irritating,
(particularly any kind of pepper)
Tomatoes & stems

Anything appearing rotten or moldy

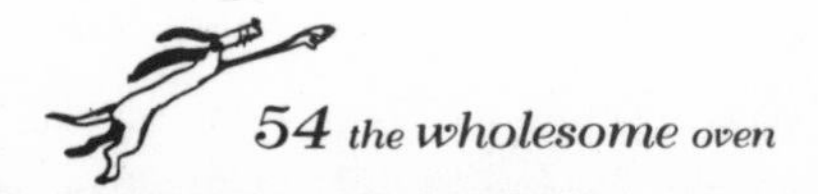

*Always avoid anything labeled simply "meat" or "meat by-products." As is now well-known, generic "meat" in animal feed can contain rotting or cancerous flesh which is too gross to sell for human consumption. It can also include road-killed animals, and euthanized zoo and companion animals (yes, still full of the killing drug). "By-products" can also include restaurant garbage (including rancid fat and oil), as well as the sweepings of a slaughterhouse floor, including blood, brains, sawdust, feathers, and feces. (Mad Cow-like prions are most often found in brain tissue and blood.).

Even "meat" with a recognizable name, like "beef" or "lamb," (or Ferdinand or Elsie) can also come from "downer" animals, who are rarely tested, and may have died from Mad Cow (BSE) or scrapie, a similar disease found in sheep. Even venison-based dog food may be deadly, because elk and deer in Colorado, Wyoming and Montana also now suffer from the Mad-Cow-like Chronic Wasting Disease (CWD). Consider this not-widely-publicized (I wonder why?) fact. Cats in Britain who have been fed Mad-Cow tainted meat, have developed a BSE-type disease. The same thing has happened to minks in North America. Nobody can be sure at this point that dogs are immune.*

The best rule of thumb is that when shopping for your dog – just as when shopping for yourself – it is critical to read the labels. If what you are reading is incomprehensible, or sounds suspicious, just say "NO."

***For further information on Mad Cow:**
www.mad-cow.org

RECOMMENDED BISCUIT INGREDIENTS

These are the ingredients I use in this book. There is general agreement that these foods are safe and nutritious for dogs. To my knowledge, allergic reactions to these are also generally rare; however, allergies can be very individualized. See pages 59-62 for more on allergens.

Brewers' yeast: This is generally considered a safe nutrient for dogs; it has a cheesy taste that they like.

Carob, powdered: Dog-treat makers have been using this for decades as a safe substitute for chocolate. For many dogs, chocolate is highly toxic. Better safe than sorry.

Flours and grains: These are the ones that I have personally used over the years and found no adverse reactions among my own dogs.

Wheat-related grains, but which do not seem to cause allergic reactions: barley, oats, oat milk, spelt, triticale.

Non-wheat-related grains: amaranth, buckwheat, millet, quinoa, rice. Some dogs are allergic to rice. I only use it in two recipes. (Cooked white basmati rice, or brown rice flour, seem most digestible. Although some holistic vets insist on cooked whole grain brown rice, I have found it too irritating to my dogs' digestive tracts.)

Fruits: These are sources of vitamins, micronutrients and phytochemicals. I suggest apples, apricots, avocados, bananas, berries, dates, peaches, pears, rosehips.

Ginger, powdered: Safe for dogs. Holistic veterinarians and doctors both suggest it for queasiness and motion sickness. If you wish to use it medicinally, check with your vet to determine dosage.

Honey: There are differing points of view on both the safety and the ethics of honey. I won't get into them here. If you don't want to use honey, use maple syrup or molasses. But don't use a grape-based sweetener.

Kelp, granules: This has a nice salty, fishy taste that dogs seem to enjoy; it is a common nutritional supplement.

Legumes: These are a source of vegetable protein. Try peas, lentils, garbanzo beans, pinto beans, kidney beans, white beans, black beans. Soybeans are very common in dog foods, which is why many dogs develop soy allergies. I only include soymilk powder in one recipe.

Maple syrup and molasses: Like us, dogs are attracted to things that aren't necessarily good for them, like large amounts of sweeteners. Commercial dog foods and treats can be packed with sugar, and the increase of diabetes in dogs is keeping pace with the increase in humans. However, there is a bit of sweetener recommended in a couple recipes, just for a special treat.

Nuts and oils: I suggest almond butter, almond milk, canola oil. Nuts can be an allergen. Peanuts become carcinogenic when moldy, but I haven't found any studies of this in dogs.

Sea salt: A little salt is okay for dogs. Many dog foods are over-salted. I only include it in a couple of recipes, in small amounts. Leave it out if you want.

Turmeric, powdered: Safe for dogs. Some holistic veterinarians prescribe it as an anti-carcinogenic supplement. If your dog has cancer, please check with a veterinary professional to determine appropriate dosages.

Vanilla: A pleasant flavor that dogs seem to enjoy. I have never heard of it being harmful, and have seen no sign of adverse reaction in my dogs.

Vegetables: These are sources of vitamins, micronutrients and phytochemicals. I suggest asparagus, bell pepper, broccoli, cabbage, carrots, cauliflower, celery, kale, lettuce, peas, potatoes, summer squash, sweet potato, taro root, winter squash, zucchini.

Water: I only give my dogs filtered water, and when I cook for them, I also use filtered water. If tap water is not safe for us, why would it be safe for them? Their systems are no better equipped than ours to handle chlorine, fluoride, toxic chemicals, and heavy metals.

ABOUT FOOD ALLERGIES

While some food allergies seem to be genetically based, many seem to result from over-consumption of a particular item. It is now very common for dogs to become allergic to the mainstay ingredients in most commercial dog foods (as well as in biscuits, training treats, and the growing category of "doggy junk food" products mimicking human junk foods, such as popcorn, candy bars, sausage, and ice cream).

The most common culprits:
Beef, pork, chicken, turkey, fish, fish oils
cow's milk products, eggs
wheat, corn, soy products
Other frequent allergens:
lamb and horse meat
rice and potatoes
beans, legumes, peanuts, tree nuts
broccoli, cabbage, carrots, cauliflower, chard, tomatoes
brewer's yeast

Since I have a very allergic dog, I have developed a layperson's understanding of how the process works in her. I will summarize it here, for those interested. Maeve was a stray pup; her mother was no doubt unhealthy. She may also have been weaned too early. Whatever her background, she clearly started life with a weak immune system, indicated by the mange she had when I took her in. Demodex mites live on virtually all dogs' skin, and many humans', but a healthy immune system suppresses their effects. This is true of staph bacteria as well. A mange or staph outbreak is a warning sign of an immune-system problem.

We cured Maeve of mange, but she developed serious dermatitis soon after. I now understand that her weak immune system was pushed to the limit by exposure to foods and plants which were major allergens for her. Her immunity broke down as her body tried to fight off all the attackers and irritants. Then the staph bacteria on her skin got out of control, causing rashes and infections all over her face, abdomen, feet and legs.

When she was about 1 year old, I heard about the Heska blood test. It revealed a long list of foods, grasses, weeds, and even trees to avoid. I immediately changed her diet to a home-cooked (fully vegetarian) one, and kept her away from the worst grasses and weeds (luckily her favorite exercise is swimming). She also received some immune-system support injections from the vet. Within a few months she was 90% healed. The deep infections on her lips and chin (which, if you're a dog, come into contact with everything) took about a year to completely clear up. I kept her on the strict diet for about 2 years; then a kibble that fit her needs finally came on the market.

Five years onwards, I had a second Heska test done, and the numerical indicators for all the allergens were by then very low, in the "safe" range. This, according to the vet, is because she has not been overexposed to them for such a long time. I still limit Maeve's contact with her main allergens, and still supplement her diet with fruits

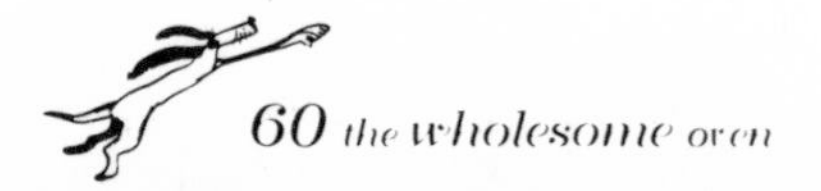

and vegetables. Now she can even consume small amounts of some allergen foods without a noticeable reaction. But if I were to put her back on a diet heavy in those foods, the allergies could come back as strongly as ever.

There are many other dietary issues affecting dogs these days. Illnesses such as cancer, obesity, diabetes, pancreatitis, Cushing's disease and kidney stones are all related to unhealthy and overprocessed commercial foods. Diet issues are further complicated by numerous congenital health problems, which result from careless inbreeding of purebreds. Many dogs do not produce the right balance of enzymes or hormones necessary to effectively process what they eat. There is no space here to discuss all these issues. See the Bibliography on page 68 for futher information.

Allergies, immunity, and diet are a very complicated field. Food allergies show up in some dogs as skin problems, and in other dogs as digestive problems, and in extreme cases as both. If you suspect your dog is allergic, my advice is this.

- Look for a vet with a holistic viewpoint.
- NEVER accept a vet's prescription for steroids. Steroids artificially suppress the immune reaction. They do not treat or cure the cause.
- DO accept antibiotics if the vet recommends them. Infections can get out of control rapidly.
- Get a Heska blood test. The vet may tell you they can be inaccurate, but Maeve's were both good indicators, and the list will give you a reasonable place to start.
- When you get the Heska results, stop feeding the allergenic culprits to your dog!

- Don't switch to a new food which is full of other major allergens (see page 59). If your dog is allergy-prone, eating lots of a new allergen will probably result in a new allergy. Keep the diet as free of allergens as possible.
- Don't feed health-undermining ingredients which can compromise the immune system (see pages 54-55).
- Supplement veterinary treatment and dietary change with a "complementary" therapy designed to support and strengthen the immune system. These include Acupuncture, Acupressure, Flower Essence Therapy, Herbalism (Chinese or Western), Homeopathy, Reiki, and Tellington Touch. Be sure to work with a qualified professional to employ any of these therapies.
- One last, important consideration. If there is a sudden allergic flare-up after years of wellbeing, do take a look at your dog's emotional state, particularly anything very upsetting – like a loved one moving away, or a companion animal dying. After being free of symptoms for five years, Maeve had a terrible whole-body relapse just two weeks after Toby's passing. Her profound grief suppressed her immune system, opening the door to the first allergen that came along. We treated her grief (with Flower Essences*), as well as her infections (with antibiotics). As her sadness faded away, her health strengthened. After about two months she was back to her usual vibrant state of health.

* For information about emotional healing with Flower Essence Therapy, see Bibliography.

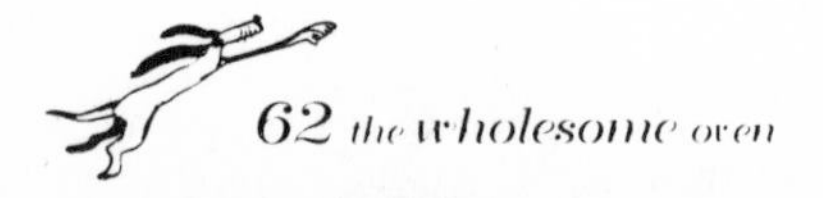

ABOUT THE DOGS

Bruce, a collie, was the first dog – and only sibling – of my childhood. He arrived on my fifth birthday. I grew up romping with him, dressing him up, and using him as a pillow while I read him dog stories out loud. Of course, he was not my personal responsibility – he was cared for and trained by my parents, who both had previous dog experience. It's a terrible idea to give a puppy to a small child, since most can't tell the difference between a sensitive live animal and a stuffed toy. It's an even worse idea to leave a puppy's upbringing to a person of any age who is not mature enough be fair, gentle, and empathetic. The puppy, like the child, needs a responsible adult setting clear rules and teaching proper social behavior.

Paul arrived to fill the too-quiet, too-empty house after Bruce's passing. We all wanted a puppy who was as different from Bruce as possible, and Paul succeeded beautifully at that. He was black, shaggy, and of completely indeterminate ancestry, with a feisty independent spirit that contrasted dramatically with Bruce's sweet, patient temperament. When my mother got a kitten that our older cat hated on sight, Paul "raised" her, and Elsa grew up thinking she was a dog – she barked, fetched, and shadowed us on walks. Paul left the world much too soon, and I still miss him.

Perdita was an elegant 20 pound black-and-white basenji mix. Along with the satiny coat, she got the basenji willfulness and joie de vivre. The non-basenji parts were the tail (straight) and the bark (loud). She came along early in our marriage, and was Karl's first-ever dog. She loved exploring old train cars and watching rehearsals at our theatre company. We named her after the castaway princess in *A Winter's Tale*. She was very much the princess, but never took on the role of castaway. She lived to be 18; during her last years her greatest enjoyment was an unrelenting intimidation of huge, genial Andrew.

Toby was our Rhodesian ridgeback mix. He was three weeks old when I rescued him from a dog pound's front desk. Raised on a bottle, he grew strong and muscular and lived to be 15 1/2. He was a champion kisser and a lifelong veggie lover. His taste for vegetarian food began early. Before his first birthday he had consumed an entire set of wood and canvas director's chairs. Books were another favorite source of roughage. In the absence of furniture or paper products, he delighted in bell peppers and carrots. We like to think his veggie preferences had a lot to do with his longevity.

Andrew was the biggest dog we've ever had. At 130 pounds he was bigger than the dogs most people have ever had. A mix of Great Dane and yellow Labrador, he united the Dane's legginess with the Lab's stocky frame. When he barked, people looked around for a sea lion. When he shook his head, the flapping of his lips sent drool flying to the upper reaches of the walls. He loved to back up to

seated people and hoist his huge rear end onto their laps. He never met a dog he didn't like. Toby, who had been aggressive with all other male dogs, fell in love with him at first sight, and they became inseparable lifelong buddies.

Tama'ehu was a small reddish terrier mix. We found him on a Kaua'i beach, so we gave him a Hawaiian name which means both "red haired child" and "sea spray child." He had ear mites, a huge abscess on one cheek, and a bullet in his side. We flew him home, thinking that once he fully recovered, it would be easy to find a good home for such a sweet little dog. An elderly couple became interested, so we took him for a visit. He analyzed the situation, and got an "Oh yeah?" expression on his face. After making sure we were all watching, he very determinedly lifted his leg on their white carpet. So we made a deal with him. We never tried to give him away again, and he never had another indoor "accident" in his life.

Maeve, our black Labrador mix, turned up one morning outside our house. Some little girls were luring her towards some busy streets. I investigated. She turned out to be a "homeless puppy" that the kids "fed and played with" sometimes. Despite mange on her face and filthy fur, she was irresistibly appealing, and I gathered her squimy little body into my arms and heart right there. She is very maternal, having devotedly mothered two foster pups as well as Ursula and Jeff. She is also extremely sensitive, and has grieved deeply at the passing of each of the four older dogs. She has such an otherworldly expression that I call her Magical Mystical Maeve.

Ursula is our chow-golden retriever mix. We found her toddling down the dusty streets of a Central Valley town. She is the family show-off. She loves getting dressed up in costumes, and insists on performing her extensive repertoire of tricks for every guest. She also has a highly-developed retrieving tendency, and will fetch just about anything requested, from clothespins to blankets (and some things not requested). She is quite catlike – loves getting up on high things, has an instinctive dislike of dogs (despite years of classes), hates to get her paws wet, chews her kibble one piece at a time, and will only cuddle when the mood strikes her.

Jeff is white, fluffy, and curly-tailed. He is bigger and stockier than the inaccurately-named "American Eskimo Dog," and we simply call him a German Spitz. We found him running loose as a four-month-old pup. When the "owners" (who had kept him tied in a grassless, treeless back yard) turned up after a week, they refused to reimburse us cash for his vet care, but were willing to trade him to us. Now he enjoys agility, has his own bed piled with pillows, and hates to stay in the yard even a moment without a human companion. He is one of the sweetest-hearted dogs I've ever known.

Seneca was our hospice-foster dog. A scarred-up old pit bull-boxer mix, he came to us from a local pound. Ill with arthritis and cancer,he was only expected to live a few days. He lived peacefully with us for another 7 1/2 months. In that time we came to love him as much as if he had been with us since puppyhood. He had clearly been a yard dog,

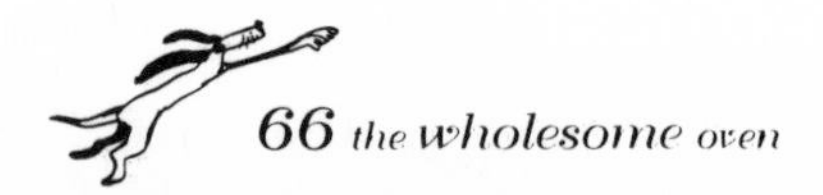

and spent much of his time here snoozing contentedly in our back garden. The last weeks or months of life can be a precious, deeply spiritual time. It should never be cut short. What a betrayal, to dump a trusting companion at a canine concentration camp (also known as Animal Control) to be executed by strangers. The Golden Rule should apply to our relations with every living being; but at the very least, we humans should treat our devoted household companions as we ourselves would want to be treated.

Winnie is the adorable Westie who lives with our good friends June and Chi-Min. She and Ursula were pups at the same time, and she became good friends with all our dogs. Terrier pups are no doubt the biggest handfuls of all. Although June was in grad school and Chi-Min had a long commute, their love and loyalty carried them through the difficult early months. They did everything they needed to do – taking Winnie to training classes, crate-training her, and enrolling her in a doggie day care facility to work off her energy safely with other lively dogs. Today she is a feisty charmer, and they can take her anywhere.

Beatrice is a toy poodle who recently left Animal Control to become the beloved companion of my dear friend Elisabeth, who works devotedly at the Fund For Animals, and who inspired me to become a vegetarian in 1974. Beatrice follows in the illustrious pawprints of Elisabeth's late companions Bix, Fideaux, and Monique. Each one was a small, adorable poodle or bichon frisé who somehow ended up at Animal Control. Each was lucky enough to go from chaos, confusion, and danger, to being the treasured companion of an extraordinarily devoted animal lover. Beatrice says, *"Don't breed, don't buy — adopt!"*

BIBLIOGRAPHY

Bach Flower Essences for Animals. Helen Graham & Gregory Vlamis. Findhorn Press, 1999.

Dogs, Diet, and Disease: An Owner's Guide to Diabetes Mellitus, Pancreatitis, Cushing's Disease, & More. Caroline D. Levin, RN. Lantern Publications, 2001.

Dr. Pitcairn's Complete Guide to Natural Health for Dogs & Cats. Richard H. Pitcairn, DVM, PhD & Susan Hubble Pitcairn. Rodale Press, Inc., 1995.

Healing Animals Naturally with Flower Essences and Intuitive Listening. Sharon Callahan. Sacred Spirit Publishing, 2001. www.anaflora.com

Pet Allergies: Remedies for an Epidemic. Alfred J. Pelchner, DVM & Martin Zucker. Dr. Goodpet Laboratories, 1986.